HOW STEM BUILT THE GREEK EMPIRE

I0808943

DONNA B. MCKINNEY

Rosen YA
New York

Published in 2020 by The Rosen Publishing Group, Inc.
29 East 21st Street, New York, NY 10010

Copyright © 2020 by The Rosen Publishing Group, Inc.

First Edition

All rights reserved. No part of this book may be reproduced in any form without permission in writing from the publisher, except by a reviewer.

Library of Congress Cataloging-in-Publication Data

Names: McKinney, Donna Bowen, author.
Title: How STEM built the Greek empire / Donna B. McKinney.
Description: New York : Rosen Publishing, 2020. | Series: How STEM built empires | Audience: Grades: 7–12. | Includes bibliographical references and index.
Identifiers: LCCN 2019013255| ISBN 9781725341449 (library bound) | ISBN 9781725341432 (pbk.)
Subjects: LCSH: Science—Greece—History—Juvenile literature. | Technology—Greece—History—Juvenile literature. | Engineering—Greece—History—Juvenile literature. | Mathematics—Greece—History—Juvenile literature. | Greece—Civilization—To 146 B.C. —Juvenile literature.
Classification: LCC Q127.G8 M35 2020 | DDC 509.38—dc23 LC record available at https://lccn.loc.gov/2019013255

Manufactured in the United States of America

On the cover: The Parthenon is the most recognizable triumph of Greek STEM achievement; it is both functional and beautiful, and it has weathered the test of time.

CONTENTS

INTRODUCTION

The word "ancient" might bring to mind something old-fashioned, dusty, or outdated. However, the ancient Greeks' contributions in science, architecture, sports, literature, drama, medicine, and mathematics paint a picture of a society that was the exact opposite—extremely forward thinking. Out of the ancient Greek culture sprang events and ideas like theater, the Olympic Games, geometry, and the idea of government by democracy—concepts that make people's lives richer and better today.

The inquisitive ancient Greeks loved to ask why questions, trying to understand how things happened or why things existed. This healthy curiosity fueled their advances in the fields of science, technology, engineering, and mathematics (STEM), many of which impact Western culture today.

Coming out of the Greek Dark Ages around 800 BCE, the ancient Greeks entered what is called the archaic period. This period brought sweeping changes to the Greek way of life. During this time, the Greeks moved away from rule by kings toward a more democratic government. The Greeks began producing their beautiful pottery, sculptures, and architecture. They also began to make and use coins for the first time. The population grew, and Greeks increased trade with neighboring countries. The archaic period ended around 500 BCE, after which Greece entered the classical period. Greek culture flourished during this era, with influential philosophers like Socrates, Plato, and Aristotle rising to fame.

Greece was organized into many city-states at this time. Each city-state—a large city and nearby small villages—had its own government, dialect, and currency. The city-state of Athens began to govern itself as a democracy during this period. It was also frequently at war with another powerful city-state: Sparta. The great military leader Alexander the Great lived during the classical period. By conquering large expanses of land, he carved out an empire that reached from Greece to India. Where his military conquests reached new lands and people, the Greek

In ancient Greece, Olympic athletes were heroes—the Greeks decorated their pottery with scenes of the athletes competing.

language, ideas, and learning spread into these regions as a matter of course.

When Alexander the Great died in 323 BCE, it ushered in the third era of ancient Greece: the Hellenistic period. The word "Hellenistic" comes from the Greek word *Hellenes*, the name the Greeks used to refer to themselves. This period lasted until 146 BCE, when the Romans conquered Greece. However, being defeated by Rome did not bring an end to the influence of the ancient Greeks. The Romans admired Greek ideas and their way of life, so the Romans continued to spread their influence, particularly in STEM fields.

The Greeks first described the earth's circumference, helped map the movement of the sun and moon, and showed that illnesses came from natural causes. They gave the world the word "mathematics," developed geometry, and showed how pulleys and levers worked. The ancient Greeks developed the catapult, which changed the way wars were waged. They built one of the seven wonders of the ancient world—the magnificent lighthouse at Alexandria, Egypt—and constructed buildings that could withstand the damage of earthquakes. The ancient Greeks gave the world significant developments across the wide-ranging fields of STEM.

CHAPTER ONE

A GLANCE AT ANCIENT GREECE: HISTORY, GEOGRAPHY, AND PEOPLE

During the Greek Dark Ages—before 800 BCE—the Greeks were farmers living in small villages. As time passed, the villagers created larger governments and collected taxes. Some villages raised armies and built walls to protect themselves. After many years, these farming villages came to be called poleis, the Greek word that translates to "city-states" in English.

People in these ancient Greek city-states spoke the same language and shared the same customs, but the individual city-states were different from each other. Some were small—with just a few hundred people living in the area of influence—while others were much larger, like Sparta, which spread across 300 square miles (777 square kilometers) of land. Each city-state governed itself.

Although the word "city" in city-state may suggest urban areas in the way the word is used today, most ancient Greeks lived and worked in the countryside. Some lived near town and traveled out to work in the fields, but farming was the backbone of Greek lifestyle and work.

This map of ancient Greece shows the Peloponnese, a large peninsula—a piece of land almost totally surrounded by water—in the southern part of the country.

CLASSES AND CITY-STATES

The ancient Greeks had distinct social classes. The people in the highest class worked in the government and education. They had slaves to do their chores. The people in the middle class worked in the trades; they were free people, like the upper class, but they did not have all the privileges of the elites. People in the lower class were often slaves who had been freed. They had fewer rights and privileges than the upper and middle classes. Slaves were the lowest class in the social order of Greek society. Typically, they were people brought to Greece from other countries. Slaves were occasionally allowed to work to earn their freedom.

Entering the archaic period around 800 BCE, the Greek city-states' economies were based on agriculture. They were no longer ruled by kings, but a small number of wealthy aristocrats held the political power in each city-state.

Every city-state wanted to possess more land, so as the population grew, the Greeks spread out in search of new territories. Greek city-states grew in number to more than 1,500 by around 600 BCE. They reached from the Mediterranean Sea across to Asia Minor and from the Black Sea down to North Africa. This included territory on the mainland and on islands.

This growth of the city-states into new lands had an important impact on Greek art, literature, and science. As people spread into new areas, they carried with them the influence of Greek ideas. The important accomplishments of the Greek scientists were given an even greater reach and impact because of the geographic growth of the city-states.

Over time, the city-states evolved from farming communities to more diverse economies. They produced and traded goods like metalwork, wine, olives, tools, and fine pottery. With the growth of trade, some Greeks accumulated great wealth. The old aristocrats who had ruled the city-states began to be replaced with new wealthy people referred to as tyrants. Some of these new leaders ruled harshly, while others brought improvements to their city-state. However, political changes were coming for the ancient Greeks. Around

500 BCE, the classical period of ancient Greece began. During this period, a system called *demokratia*, or "rule by the people," began. This was the first appearance of the democratic system of government.

MOUNTAINS AND MORE

Greece—one of the most mountainous countries in Europe—is situated on both the mainland of southeastern Europe and spread over a thousand islands. The Greek people lived in mountainous lands touched by the sea and believed their

As a country of mountains surrounded by water, since ancient times the terrain and sea have had an impact on the ways Greece grew and flourished.

many gods lived atop Mount Olympus, the tallest mountain in Greece. Running north to south across mainland Greece, the Pindus Mountains are sometimes called the spine of Greece.

The Mediterranean region where the ancient Greeks lived is called the Aegean. The Aegean Sea provided food for the people and gave them a way to travel by boat from one city to another, as well as easy access to other islands and countries.

This combination of mountain and sea led to Greece being a decentralized country, hence the emergence of multiple independent city-states. The geography also contributed to the agriculture and diet of the region. The hills provided good conditions for growing grapes and olives, used to make wine and oil, respectively. The coastal areas had soil that was good for growing barley and wheat. Seafood became an important part of the diet for a country sitting along the coast and dotted with islands. Because of the Greeks' location around the sea, trade and travel were easy for the Greeks. Fisherman, sailors, and merchants who traded goods all found work because of the nearby sea. The natural geography of mountains and seas also carved out several distinct regions among the ancient Greeks:

- The Peloponnese is a large peninsula on the southern end of mainland Greece. A tiny strip of land called the Isthmus of Corinth connects the peninsula to the mainland. Some of the bigger city-states—like Sparta, Corinth, and Argos—were located in the Peloponnese.

- Central Greece is located north of the Peloponnese peninsula. The city-state of Athens is in central Greece. The famous Greek marble is also found is this region, which is even more mountainous than northern Greece.
- Northern Greece is home to Mount Olympus and the Pindus mountain range. Within northern Greece, there are three subregions: Thessaly, Epirus, and Macedonia.
- The many Greek islands were grouped by their location, including the North Aegean islands, the Dodecanese islands, and the Cyclades islands. Crete and Rhodes were among the largest of the islands.

FOR THE LOVE OF SPORTS

The ancient Greeks loved sports, so the Olympic Games were an important part of their cultural identity. The earliest records show that the games were held in 776 BCE, although they may have started years earlier. The events were created to honor the Greek gods—especially Zeus, the most important member of the pantheon. Only the men attended or competed in those early Olympic Games, at which records show that there was just one event: a footrace of about 200 yards (183 meters).

For a people divided into city-states and settlements—separated by mountains and the sea—the games united the Greeks in their worship of Zeus and their love of sports. Even when the city-states were at war with each other, they called a truce and stopped fighting when the time for the games arrived. People came together from across the far-flung reaches of the Greek empire for the athletes to compete and the spectators to cheer them on to victory.

After the Romans conquered the Greeks, they banned the Olympics in 393 CE. Centuries passed until the Olympic Games were revived in 1896. The games have been played every four years since then (except during the war years of World War I and World War II).

The ancient Greeks loved athletics, focusing on individual sports like running, javelin throwing, wrestling, long jump, and chariot racing.

GREEK INFLUENCE SPREADS TO NEW LANDS

As the ancient Greeks spread out in search of foreign lands, they built colonies across the region of the Mediterranean and Black Sea. As Greek people traveled and settled, they brought Greek culture and learning with them. With the settlement of these colonies, Greek influence spread into northern Africa, France, Spain, and Turkey. Under the military leadership of Alexander the Great, the Greeks spread their empire into Egypt and India. Their amazing achievements in science, technology, and medicine came about despite the fact that they were often busy fighting wars.

The way the ancient Greeks governed themselves contributed to the rise of STEM, as well as the advancement of literature and philosophy. The city-states were generally prosperous, so the Greeks were able to put significant resources toward education, learning, and research. Out of their love of learning, outstanding teachers arose, attracting students to their schools, where they explored ideas about nature, philosophy, and mathematics. Before long, the Greeks were making significant contributions to the body of knowledge surrounding math, astronomy, physics, medicine, and nature.

CHAPTER TWO

THE GREEKS GAVE US GEOMETRY: ANCIENT GREEKS AND MATHEMATICS

With many discoveries in STEM fields, researchers in one generation build on the work of earlier generations. Building on the math of the Egyptian, Babylonian, and Chaldean cultures, the Greeks applied this borrowed knowledge and made their own breakthroughs. The Greeks' discoveries in mathematics provided the foundation for generations of mathematicians and scientists who followed them.

The Greeks applied their math advancements to solve problems in other fields, like astronomy and engineering. Where the Egyptians and Babylonians depended on their observations of the natural world to help them solve problems, the Greeks took a groundbreaking new approach to their study of math. Instead of relying on observations, the Greeks developed the idea of the proof—where they used logical steps to either prove or disprove theorems. This approach gave the Greeks the tools to make incredible advancements in math that would influence mathematicians for centuries to come.

BORROWED MATH

In math, the Greeks looked to their neighbors and borrowed from the even more ancient cultures of the Egyptians, the Babylonians, and the Chaldeans. Armed with the influences of those other cultures, the Greeks pushed forward to even greater accomplishments with numbers. Around 600 BCE, the Greeks began trading with the Egyptians. This trade allowed important Greek mathematicians—like Thales and Pythagoras—to visit Egypt, further exchanging ideas and knowledge.

The Egyptians used mathematics for practical applications, like building the great pyramids, so they knew that a right triangle is one in which the sides are in a 3:4:5 ratio—and they used this in their engineering work. The Greek mathematician Pythagoras took that 3:4:5 ratio triangle, and from that he created the well-known math formula that is called the Pythagorean theorem.

Hundreds of years later, when Alexander the Great conquered the lands east of Greece, the Greeks soaked up the math and astronomy knowledge of Babylonia and Chaldea. From these eastern cultures, the Greeks adopted the use of a numeral system based on sixty, and this allowed them to divide a circle into 360 degrees—another major breakthrough.

THALES OF MILETUS

Thales, who lived around 620 to 546 BCE, is one of the earliest Western philosophers and mathematicians. He lived in Miletus—a Greek colony in the western part of present-day Turkey—around 585 BCE. He studied a wide range of topics, ranging from astronomy to electricity to geometry. It would be impossible to overstate his importance in the history of STEM advancement.

Thales is thought to be the first of the Greek philosophers, and many view him as the father of science. Thales may have studied in Egypt or Babylon, although this is not known for sure. It seems likely that he at least studied Egyptian astronomy, math, and mythology..

Thales contributed to the development of geometry, introducing the idea of abstract geometry. He found practical uses for geometry, using it to calculate things like the distance between ships and the shoreline or the height of the pyramids.

None of Thales's writings exist today. All that is known about him comes through other people, including Aristotle, who wrote about his work hundreds of years later. Thales focused on the idea of First Cause—asking what everything else comes from. He concluded the First Cause was water, and that it was the essence of all other things. His argument for water being the First Cause was the first time a scientist had tried to explain nature and the physical world without referring to the supernatural power of a god. As Thales

Thales of Miletus is considered to be the "father of science" because of his contributions in so many areas of science and math.

shared his ideas, the shift from superstitions to science began.

People in the ancient world also believed that earthquakes were caused by some kind of divine fury of the gods. In contrast, Thales proposed that the earth floated on a large body of water and that earthquakes were caused by water turbulence. While scientifically he was wrong, Thales again was helping to move people away from looking to the supernatural for explanations of the natural world.

Thales founded the Milesian School, at which young men studied. Anaximander, an accomplished philosopher in his own right, was one of Thales's most famous students at the school.

PYTHAGORAS

Pythagoras was a philosopher and mathematician born on the island of Samos about 570 BCE. Many recognize him as the first mathematician. It is believed that he studied with both Thales and Anaximander, some of the earliest Western philosophers.

Math students today probably recognize the formula $a^2 + b^2 = c^2$. In this familiar geometric argument, Pythagoras described the relationship between the sides of a right triangle. Although Pythagoras proposed many mathematical theorems, the Pythagorean theorem is the most enduring and best known in modern math.

Pythagoras believed that scientists could understand and explain the physical world through

Pythagoras, a mathematician, believed the physical world could be understood and explained through math. Today, math students learn about his Pythagorean theorem, which describes the relationship of the sides of right triangles.

math. For example, he proposed the idea that the planets and stars moved in ways that could be explained through mathematical formulas. He had followers, called Pythagoreans, who studied numbers and believed that the soul goes through cycles of rebirth. They worshipped the god Apollo. Pythagoras influenced other philosophers, including Plato. Besides his influence in mathematics, Pythagoras also studied music theory, medicine, and astronomy.

EUCLID OF ALEXANDRIA

Euclid, a Greek mathematician who lived around 300 BCE, is often called the father of geometry. The exact years of his birth and death are not known, but he lived in Alexandria, Egypt, and worked in the great Library of Alexandria. He also started the Alexandrian School of Mathematics, where others came to learn all that Euclid knew and could teach about math. In the field of mathematics, Euclid's influence is easily seen in the things that bear his name today: Euclidean geometry, Euclid numbers, and the Euclidean algorithm.

Euclid collected all that the Greeks knew about math up to that time and wrote a math textbook called the *Elements*. Euclid included the work of mathematicians such as Pythagoras, Hippocrates, Theudius, Teaetetus, and Eudoxus, bringing all that math knowledge to bear in a cohesive, singular work. This large work consists of thirteen volumes covering arithmetic, geometry, and number theory. Through time, the *Elements* has continued to be an important math textbook, if for nothing else than its impressive historical significance.

The principles Euclid developed are the foundation of Euclidean geometry, which is still in use today. This type of geometry uses a small set of axioms instead of long theorems. Across a long period of time, Euclid's geometry was the only geometry that existed. In modern times, it is Euclidean geometry that is typically taught in high schools. It is also sometimes called parabolic or flat geometry. It is still considered to be as relevant in the study of math today as it was when he first developed it more than two thousand years ago.

Euclid did not devote all his time to studying math, however. He also wrote about a dozen other important books on topics as diverse as optics and music. About half of the works Euclid wrote are lost, and all that is known of those works today are clues left by other writers who mention them. Nonetheless, his contributions to the long history of Greek mathematical advancement are far-reaching.

ARCHIMEDES

Archimedes's father was an astronomer, so perhaps it was no surprise when he followed in his father's footstep in the study of the heavens. Archimedes (around 287–212 BCE), however, was also an inventor and a mathematician. When he was about ten years old, Archimedes left the Greek city of Syracuse to attend school in Alexandria. There, he studied math in

Archimedes's important work in mathematics laid the foundations for modern calculus and mathematical physics.

the school Euclid started. Archimedes is considered to be one of the greatest mathematicians who ever lived.

In his study of mathematics, Archimedes was far ahead of his time. Some of the complicated equations he developed would not be fully understood until 1,800 years later, when Sir Isaac Newton extended the work Archimedes began. He laid the foundations for modern calculus and mathematical physics.

Archimedes developed the method of exhaustion to calculate the area inside a circle. He computed the area and volume of the sphere. He developed the concept of actual infinity. He was the first mathematician to give an accurate estimate for the value of pi (π), one of math's most important numbers. This is considered by many to be one of Archimedes's greatest contributions to mathematics.

Archimedes also developed the discipline of statics, a part of applied mathematics that deals with loads that do not move, such as those in bridges or buildings. He also examined how objects float in liquids and wrote about buoyancy. Archimedes discovered a law related to buoyancy, called Archimedes's principle, still used in the field of fluid mechanics today.

Archimedes's remarkable contributions to the field of mathematics are expansive. He stands alongside later mathematicians—including Isaac Newton and Carl Friedrich Gauss—in the importance of his impact on modern-day mathematical studies.

HIPPARCHUS OF NICEA

Hipparchus was born in Nicea, in modern-day Turkey, around 190 BCE. He was an astronomer, geographer, and mathematician. Hipparchus invented trigonometry, which simplifies the math of triangles. Scientists were then able to use trigonometry to simplify the calculations they used in their astronomy studies. Hipparchus was also famous for his contributions to astronomy, a major field of STEM advancement for the Greeks.

CHAPTER THREE

BEYOND THE WORLD: ANCIENT GREEKS AND ASTRONOMY

The Greeks approached astronomy by building on the foundation laid by the Babylonians. For year the Babylonians had collected very detailed observations of the planets and the stars. They applied math to their observations, allowing them to predict certain events in the sky. Despite an obvious interest in the heavens, however, the Babylonians did not explore the underlying causes for what they observed.

The Greek astronomers took those observations by the Babylonians (and those of other ancient societies) and carried the study of astronomy to remarkable new heights. They built models that dissected how the universe worked. The Greek astronomer Aristarchus correctly proposed the idea that the sun—not the earth—was the center of the solar system. It took centuries for the rest of the world to catch up with his theory.

The Greek astronomers believed their gods controlled the sun, stars, and planets. Their earliest astronomers were priests with strong ties to astrolog

RELIGION AND THE ANCIENT GREEKS

In ancient Greece, religion touched every area of most people's lives. The gods' activities were used to explain anything Greeks observed in daily life and nature. With the rise of influential and incredibly intelligent philosophers like Aristotle, Archimedes, Socrates, and Plato, the Greeks first began to look at their world based on reason, rather than superstitions. They started to investigate questions using science to find answers in the natural world and through the use of logic. These earliest shifts from religion to science created a great clash of ideas between the philosophers and scientists who proposed the new ideas and the rest of their world.

The ancient Greeks believed in many gods, male and female, who controlled every part of their lives. Nonetheless, STEM innovators led the charge in convincing Greek society to respect logic.

However, as time passed, the Greeks slowly moved away from their superstitions. They began to apply physical causes to their understanding of the universe.

Not many of the Greek writings from before the fourth century BCE survive, so today's knowledge of those earliest astronomers is based mainly on what Aristotle later wrote about them. Despite this lack of written records, the ancient Greeks' contributions to astronomy would stand for centuries. Some of the concepts of modern astronomy are built on ideas first suggested by the ancient Greeks.

THALES AND PYTHAGORAS

Thales, highly famous for his contributions to math, was also an astronomer. He discovered the constellation Ursa Minor, commonly called the Little Dipper. He also built a simple telescope, as he wanted to better observe the stars.

Thales is reported to have predicted a solar eclipse in 585 BCE—a prediction that ended the war between the Lydians and the Medes, who had been fighting for years. The eclipse surprised the soldiers so much that they stopped fighting and declared a truce. Modern-day scientists have questioned how a scientist from that time could have possibly predicted an eclipse, so some wonder if this is simply a legend.

Pythagoras—another Greek extremely well known for his contributions to math—was also an accomplished astronomer. He was the first to suggest that the earth was a sphere. He applied math to explain the movement of the planets, sun, stars,

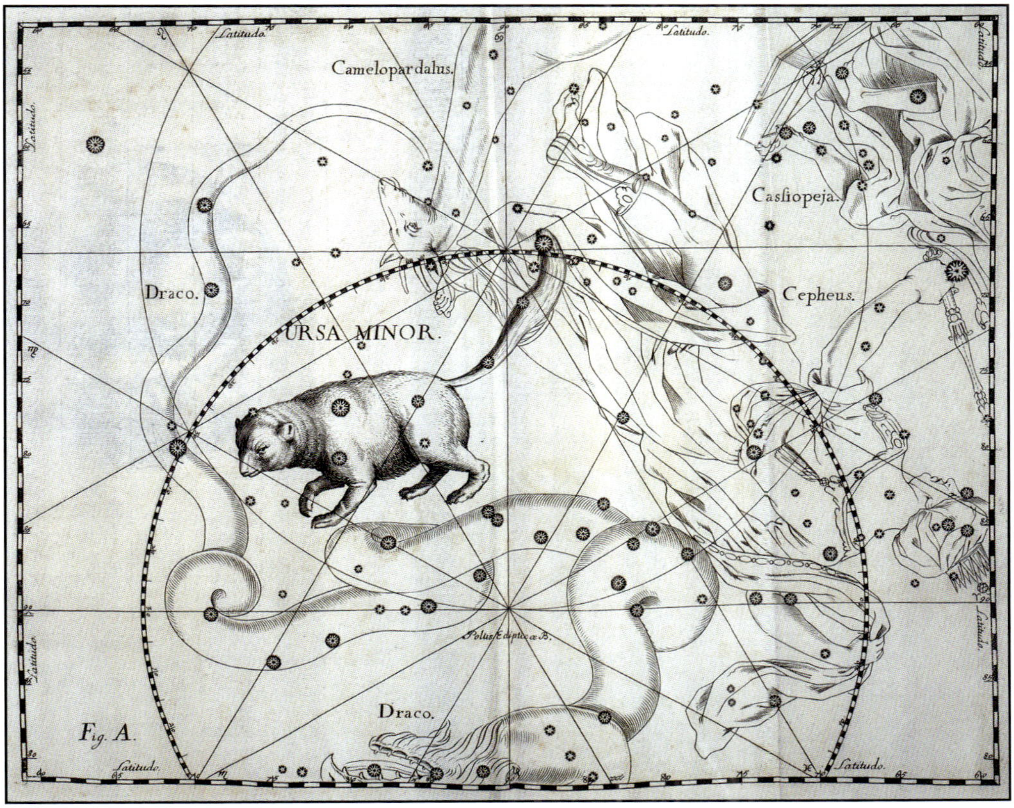

The constellation Ursa Minor—its more familiar name being the Little Dipper—was discovered by Thales, the famous Greek mathematician and astronomer.

and moon. He was the first to understand that the planet Venus is both the morning and evening star.

ANAXIMANDER OF MILETUS

Anaximander (around 610–546 BCE) is famous for being one of the first cosmologists—a person who describes the origin and form of the universe. He is also believed to be one of the earliest Western philosophers. At the time Anaximander lived, the Babylonians and Egyptians practiced observational astronomy. Their studies consisted of what they observed with the naked eye, such as the ways the moon, stars, and planets moved across the sky. By contrast, Anaximander's achievements lie in the field of speculative astronomy. He was making educated guesses about things that could not be seen. For example, he believed the planets made full circles, passing beneath the earth. He was the first to suggest the revolutionary concept that the earth does not rest on top of something else, such as an ocean. He hypothesized that the earth is its own body.

He was the first Greek scientist to believe that the planets and stars existed behind each other and that they were not all the same distance from the earth. He understood that space had depth. Anaximander also believed that the earth was sitting still, as the center of everything. He believed in a geocentric model of the universe, and while this idea was incorrect, his arguments were widely believed until 1543 CE, when Nicolaus Copernicus definitively proved Anaximander's idea to be false.

Anaximander created the earliest mechanical models of the universe. He drew what is believed to be one of the earliest maps of the world. He also developed a kind of sundial called the gnomon. He proposed the idea that lightning storms are caused by air disturbances, not by angry gods—again, shifting scientific understanding away from superstition.

Where Anaximander's teacher, Thales, believed the First Cause of everything was water, Anaximander viewed water as just one of the natural elements. He believed, therefore, that the First Cause came from something else other than water—something Anaximander called *apeiron*, or "the boundless." He argued that this was the cause or origin of everything that exists.

EUDOXUS OF CNIDUS

Like many learned Greeks, Eudoxus (around 395–342 BCE) possessed a well-rounded education. At different times, he studied math, astronomy, medicine, philosophy, and rhetoric.

As an astronomer, Eudoxus combined his understanding of astronomy and geometry to develop the astronomical globe. He was also responsible for producing the first sophisticated model displaying the movement of the planets. From his observatory in Cnidus—in Asia Minor—Eudoxus identified the constellations.

In mathematics, Eudoxus made important contributions in number theory and geometry. Archimedes would build on Eudoxus's work in math, and the two of them are credited with laying some essential groundwork toward the development of calculus.

As with other ancient Greek scientists, only fragments of Eudoxus's writings exist today. Eudoxus was honored by modern-day astronomers with craters on Mars and the Moon named after him. An algebraic curve, called the Kampyle of Eudoxus, is also named after him.

ARISTARCHUS OF SAMOS

Aristarchus was a mathematician and astronomer born on the island of Samos in the Aegean Sea. He lived about the same time as Archimedes and Erathosthenes, estimated at around 310 to 230 BCE.

Aristarchus proposed the startling idea that the sun—not the earth—was the center of the universe, concluding that the earth revolves around the sun. "Heliocentric" is the word now used to describe the sun-centered view of the universe that Aristarchus proposed. This was directly and completely opposed to the much more common geocentric model.

Many Greek scientists rejected Aristarchus's sun-centered model, fixated on their geocentric view. The only other astronomer who is known to have supported Aristarchus's sun-centered view was Seleucus, who lived about a hundred years later. It was not until nearly 1,800 years later that Copernicus, Johannes Kepler, and Isaac Newton began to explain the heliocentric model, using the laws of physics and advanced mathematics to support it.

In addition to his ideas about a heliocentric universe, Aristarchus also proposed the new idea that the stars are positioned a great distance from the earth—an idea that, in modern times, is common knowledge, but back in his era, was revolutionary. Unfortunately, like many other Greek scientists, most

Aristarchus and Hipparchus were two of ancient Greece's most highly respected astronomers. Each proposed groundbreaking ideas.

of Aristarchus's writings have been lost; what is known about him was gathered through the writings of others. Today, Aristarchus has a crater on the moon named after him.

ERATOSTHENES

Eratosthenes of Cyrene (around 276–194 BCE) wrote books on a range of topics, including philosophy, mathematics, history, astronomy, and poetry. His knowledge in so many varied fields likely helped him get the invitation to serve as the director of the great Library of Alexandria, a major research center of that time.

One of Eratosthenes's major contributions to STEM history was his measurement of the earth. He studied how the sun cast shadows, especially at the time of the summer solstice in June, and noticed something very important. In Alexandria—where he lived—in June, the sun cast shadows at noon, shining down on buildings even when it was directly overhead. However, in the city of Syene, south of Alexandria, the sun cast no shadows when it was directly overhead on the solstice. By comparing these two things, Eratosthenes understood something others had missed. He believed that the earth was a sphere, and using geometry, he calculated the circumference of the earth.

In addition to calculating the earth's circumference, Eratosthenes measured the distance between the earth and the moon and catalogued the stars. He also measured the distance from the earth to the sun, a measurement called the astronomical unit. His world map—extremely detailed and meticulously calculated—was a first at that time. He included the equator and lines, similar to the latitude and longitude lines in standard use today. Eratosthenes came up with the word "geography"—the study of the earth—and pioneered that field of study.

HIPPARCHUS

Hipparchus, who rose to fame as a mathematician, was also an astronomer and a geographer. Born in Nicea, some consider him to be the greatest of the

HIPPARCHUS

THE FOUNDER OF THE SCIENCE OF ASTRONOMY, AND THE MOST FAMOUS OF THE GREEK ASTRONOMERS.

Hipparchus, a mathematician and astronomer, catalogued the position of the stars long before the telescope was invented.

Greek astronomers. There is a crater on the moon named after him, in honor of his accomplishments and contributions to STEM. Hipparchus was the first to develop accurate models for the way the sun and moon moved. He assembled the first extensive catalogue of the stars, noting the positions of more than one thousand stars.

In one of his most important achievements, Hipparchus is credited with discovering the precession. This is a name for the ways the stars move across the sky from the viewpoint of Earth. For example, over thousands of years, the North Star's place in the sky changes just slightly. Hipparchus estimated that the precession of the equinoxes (the two times a year when the sun is directly over the equator) was thirty-six seconds each year. This calculation was just a little bit short—modern calculations reveal that it is fifty seconds.

It is also likely that Hipparchus was the first astronomer to accurately predict when solar eclipses would occur. He used trigonometry to make these calculations. Hipparchus also built astronomical instruments to observe the planets and stars. During his life, Hipparchus wrote at least a dozen books, but little of his work remains.

CHAPTER FOUR

SO MUCH SCIENCE: ANCIENT GREEKS AND PHYSICS, ZOOLOGY, AND BOTANY

Modern scientists tend to choose a particular discipline to study, such as physics, astronomy, mathematics, or biology. Often, they specialize within the discipline, focusing on narrow, specific areas of interest. However, the Greeks did not specialize in a similar manner. The amount of information and knowledge available to them was much more limited than it is today, so Greek scientists—in a rush to question and define the world—often studied broadly, becoming experts across several disciplines. It was not unusual for Greeks to study botany and physics, math and zoology, or physics and medicine. An educated Greek man might be equally skilled in discussing both philosophy and the sciences.

EMPEDOCLES

Empedocles (around 490–430 BCE) was born on the island of Sicily into a wealthy family. He was educated by the Pythagoreans, who believed that numbers had supreme importance in the universe. He was a

Empedocles invented the idea that the world consists of four elements—earth, air, fire, and water. These are known today as the classical elements.

philosopher and a poet who explored wide-ranging questions in the sciences.

Empedocles was the first to believe that everything is made up of four essential elements: fire, air, water, and earth. Aristotle himself supported Empedocles's idea of the four elements, and this was the accepted belief until the time of the Renaissance, which began around 1300 CE.

Empedocles was the first to propose the idea of natural selection, which was the theory that—over time—the organisms best suited to their environment survive and pass their adaptations on to later generations. Much later, this idea was expanded upon and popularized by Charles Darwin.

Empedocles also proposed that the speed of light is finite. He concluded that the light of the sun does not reach earth instantly; its speed has limits. Although this idea was unpopular at the time, his understanding was correct. It was hundreds of years later when scientists began to better understand that the speed of light is finite and define its properties using mathematics. Empedocles also thought that a person's senses were limited by physical form. He correctly hypothesized that there are things the senses simply do not allow a human to observe.

DEMOCRITUS

Democritus was born around 460 BCE, living around the same time as Socrates. Some called him the Laughing Philosopher because he was known for his cheerfulness. Democritus is famous for developing an early atomic theory of the universe. Atomic theory is the concept that everything in the universe is made of tiny bits of matter that cannot be divided. Democritus called these bits of matter *atoma*, which means "indivisible units." The English word "atom" comes from the Greek atoma. Democritus understood that these atoms fit together like a puzzle, and he believed that atoms were the smallest pieces of matter. Scientists in later generations came to understand that atoms are not the smallest pieces of matter, but they are among them.

Democritus did not limit his studies to atoms, however. He also explored astronomy, physics, zoology, botany, and medicine. He was the first to suggest that the Milky Way was actually the light from many distant stars. He also proposed the forward-thinking idea that there could be life on other planets, though this is still an unproven claim.

ARISTOTLE

Aristotle (384–322 BCE) is one of the most respected philosophers of all time, Greek or otherwise. He conducted pioneering work in so many areas of study related to the sciences and the animal world. At age seventeen, he became a student of the philosopher Plato at the Academy. In most classrooms today, specific subjects are taught by specific teachers. At Plato's Academy, however, students and teachers would pose questions to each other and then respond in a debate format. As new questions arose in the course of the conversations—called dialogues—they would debate those questions as well.

Before Aristotle, most philosophers and scientists spent their time thinking about the world around them. Aristotle practiced a new approach: he observed his world, taking notes and keeping detailed records of what he saw. To learn about anatomy, he dissected animals. To learn about biology, he drew different parts of animals and then classified animals into different groups. Aristotle's ways of learning about the world through physical observations set the stage for what would—much later—be called the scientific method. This method, which involves controlled experiments, is the way that scientists study the world today.

Aristotle later became the tutor for the son of Macedonia's king—who grew up to become Alexander the Great. As a military leader in his adult life, Alexander conquered wide regions of land and spread the Greek ideas—taught to him by Aristotle during childhood. After he tutored Alexander, Aristotle returned to Athens. There, he started his own school at the Lyceum, called the Peripatetic School.

Aristotle was the first to classify animals into groups with similar characteristics. He presented this information in his book *History of Animals*. He also wrote *Movement of Animals* and *Progression of Animals*. Because of these important works about the different species of the animal kingdom, Aristotle is often called the father of zoology.

During the time he lived, people called him "the man who knew everything." Plato called him "The Mind." Later, he was referred to simply as "The Philosopher" or "The Master." He was skilled in wide-ranging subjects, such as biology, medicine, math,

Aristotle was so influential that he tutored Alexander, the son of the king of Macedonia—he would later be known as Alexander the Great.

botany, literature, logic, politics, physics, logic, and theater. One way that Aristotle lived up to these prestigious nicknames was in organizing the different areas of knowledge at that time. He studied the general knowledge of his day and classified it into categories such as psychology, physics, metaphysics, logic, rhetoric, and poetics. By organizing the knowledge in this way, Aristotle laid the foundation for most of the sciences studied around the world today. He wrote about two hundred books on a variety of topics, including psychology, science, ethics, politics, rhetoric, and reasoning. Only about thirty of his books remain today. Plato, Socrates, and Aristotle are often regarded as the greatest Greek philosophers.

THE COUNTLESS CONTRIBUTIONS OF GREEK PHILOSOPHY

The ancient Greeks developed a new way of thinking about the world. They called these new ideas *philosophia*, which means "love of wisdom." The Greeks asked big questions about how the universe worked and how people should behave. Then they spent time studying and trying to unravel these big questions. Their influence continues to have an impact on philosophy

(continued on the next page)

(continued from the previous page)

Plato founded and taught at the first school in the world for philosophers—it was called the Academy, and it was located in the city-state of Athens.

and politics today. These are three of the best-known and most influential philosophers:

- Socrates challenged his students to question everything and then use discussion and debate to explore new ideas. This form of questions and answers is called the Socratic method. He asked questions about good and evil, justice, and how a society should work. None of Socrates's written works remain today, so most of what is known of Socrates comes from his student Plato.

- Plato was a student of Socrates, but in many ways, he grew into his equal. He recorded Socrates's ideas and then later wrote his own ideas. Much of his writing takes the form of conversations called dialogues. *The Republic* is Plato's most famous work. He also wrote books on a variety of topics, including the best kind of government. Plato was responsible for founding the world's first school for philosophers in Athens. It was called the Academy, and Plato taught there for decades.

- Aristotle was a student of Plato, and he is perhaps the most famous of all ancient Greeks. He wrote about topics like science, poetry, and the animal world. Aristotle founded a school called the Lyceum. He tutored Alexander the Great. Later, as Alexander the Great conquered large parts of Asia, the Greek way of thinking he learned as Aristotle's student was spread to lands far from Greece.

THEOPHRASTUS

Theophrastus (around 372–287 BCE) was born on the Greek island of Lesbos, near modern-day Turkey. His name at birth was Tyrtamus, but it was later changed—by Aristotle—to Theophrastus because he was such a gifted speaker. In Greek, *theos* means "divine" and *phrazein* means "to speak." As a young person, Theophrastus went to Athens to study at Plato's school. While in Athens, he met Aristotle. After Plato's death, Theophrastus began to study, work, and travel with Aristotle.

When Aristotle was forced to leave Athens, he appointed Theophrastus to succeed him as the head of his school at the Lyceum, and Aristotle gave him his library of books. Theophrastus headed the Lyceum for more than three decades. The school flourished under his leadership, and grew to house more than two thousand students.

Like Aristotle, Theophrastus studied and wrote about wide-ranging topics, including religion, ethics, math, meteorology, physics, and natural history. Also like Aristotle, he focused a lot of his attention on the study of nature. His two books—*History of Plants* and *On the Causes of Plants*—are considered to be the most significant books about the botanical sciences in ancient times and remained heavily influential until the Middle Ages. Because of his important work with plants, Theophrastus is often called the father of botany.

Theophrastus studied plants, classifying more than five hundred of them as trees, shrubs, undershrubs, or herbs. In his writings, he described how seeds germinate. He understood the anatomy of flowers, noting that some flowers have petals while others do not. Besides botany, Theophrastus also wrote about human physiology and zoology.

CHAPTER FIVE

ADVANCING HUMAN HEALTH: ANCIENT GREEKS AND MEDICINE

During the hundreds of years that spanned the time of the ancient Greeks, scientists' views on sickness and healing shifted. In the fourth and fifth century BCE, the Greeks called on their gods and goddesses to intervene when they were sick. Illness was frequently seen as punishment from the gods, and healing was the opposite: a divine gift. Doctors treated their patients with herbal medicines and simple surgeries. Any illnesses or physical pain were blamed on the gods. Doctors' main job was to pray to the gods that the patients would be healed.

RELIGION MEETS SCIENCE

As time passed, Greek doctors started to look for actual physical causes when illness struck, instead of assuming superstitious, otherworldly causes. Doctors' interest in studying the human body grew, and they started to explore which treatments worked—and, importantly, which did not—with specific illnesses.

When Hippocrates began practicing medical science, the Greeks' view toward medicine and illness continued shifting toward a focus on physical causes for illness. The Greeks' interest in the physical causes of sickness, however, does not mean they completely turned away from the spiritual world. Asclepius, the son of Apollo, was the Greek god of medicine and healing. The Greek people often called on Asclepius to heal their ailments, seeking medical advice from Asclepius through dreams.

The Greeks' military activities also led to changes in medical care. Whether it was fighting among the Greek city-states or battles in the conquest of new lands, warfare always resulted in injuries and

This image shows Asclepius, the Greek god of medicine and healing; the ancient Greeks prayed to him for healing when they were sick.

casualties. The Greeks gained practical, hands-on medical knowledge as they worked to improve their treatment of the wounds that inevitably occurred on the battlefield.

GREEK MEDICAL SCHOOLS

Ancient records give clues about the earliest Greek medical school, located at Cnidus around 700 BCE. There was also a medical school located at Kos (also spelled Cos). These medical schools did not consist of a campus with buildings and laboratories, like modern-day medical schools. Rather, the Greeks' medical schools sprang up where there was an influential medical teacher who had followers. Often, these teachers would work with their students in a kind of master-apprentice relationship, with the teacher passing on his medical skills and knowledge to the younger students.

Alcmaeon of Croton lived sometime around 500 BCE and worked in the school at Cnidus. He seems to have been the first person to propose the intriguing idea that illnesses could be traced to internal causes. He even suggested the forward-thinking idea that factors like nutrition, lifestyle, and environmental causes had an impact on health and sickness.

Hippocrates, regarded as the father of medicine, founded the medical school at Kos. Much of the medicine that came out of these schools at Cnidus and Kos could trace its roots to even more ancient civilizations in Egypt and Mesopotamia, after which it was honed and improved by Greek teachers and students.

HOMER'S *ILIAD*

Homer's *Iliad* is an epic poem written about 800 BCE. It tells the story of the Trojan War between the Achaeans (the Greeks) and the Trojans. From a literary viewpoint, the *Iliad* is one of the most important classical works and is widely taught throughout the world. From a medical viewpoint, the *Iliad* is interesting because of its graphic descriptions of the soldiers' injuries and the medical treatment they received. Head wounds were almost always fatal for the soldiers, whereas wounds on other parts of the body were often treatable. The warrior heroes in the poem

The *Illiad*, an ancient Greek epic about the Trojan War, offers clues about the medical skills of the ancient Greeks in its detailed descriptions of the care of the wounded soldiers. This marble tablet shows a Trojan War battle.

provided medical care to wounded soldiers, and the text describes care for the wounded based on medical science, not on religious superstitions.

The epic also gives detailed descriptions of muscles, organs, and bones. It provides some clues about the medical treatments available at that time. In addition to the warriors, like Achilles, who possessed medical knowledge, the poem tells of two physicians—Machaon and Podalirius—who cared for the wounded soldiers. Wounds were commonly treated with washing and herbal medicines.

Homer's descriptions of the human body and the treatment of wounds are so detailed that some wonder if he may have been a doctor or have come from a family of doctors. Although the *Iliad* was not a medical text, it does give an interesting literary snapshot into the medical knowledge available among the Greeks before Hippocrates. This poem provides some of the earliest information available about Greek medicine, as countless other documents and texts have been lost to time.

HIPPOCRATES OF KOS

Hippocrates, born on the Greek island of Kos around 460 BCE, is the best known of the ancient Greek doctors. Today, he is called the father of medicine for his incredible contributions to the fields of medical science and biology. He lived about the same time as Socrates and Plato, and his work—though in different subjects—was no less important to the advancement of Greek STEM.

The Greek doctor Hippocrates, called the father of medicine, was among the first to use science to connect diseases to physical causes.

Hippocrates had a strong foundation to build on when he began studying medicine. Both his father and grandfather practiced healing arts. He learned medicine from his father, and eventually, he opened an important medical school in Kos to teach what he had learned to his sons and other students. He also wrote medical textbooks. Many doctors today swear and follow the Hippocratic oath, which is credited to Hippocrates. The oath is a statement of medical ethics that guides doctors' practice of medicine.

Doctors in ancient Greece had a difficult time learning about the human body and anatomy because dissecting a human body was simply not done. Because medical science was not very advanced, it was rare for a doctor to perform an invasive procedure like surgery. Instead, patients were typically given simple treatments like rest and cleaning wounds with water. Doctors were able to do procedures like setting broken bones or draining the lungs of a person who had pneumonia. They could also perform amputations, mainly done for battlefield wounds or accidents. However, the dangers of true surgery meant that Greek doctors chose it only as a last resort.

Hippocrates believed that people's lifestyle, diet, and environment contributed to their health. He was also one of the first doctors to suggest the idea of patient confidentiality, meaning that doctors should not tell others what their patients tell them. Hippocrates stressed the importance of cleanliness in a medical environment and treating the patient with kindness.

He kept medical records for each patient that helped him understand the course an illness might take.

While many of Hippocrates's beliefs advanced the field of medicine, some of what he taught was in error. For example, he believed—incorrectly—in the concept of humors, which were the four elements of the body: blood, phlegm, black bile, and yellow bile. He believed that the humors being imbalanced was responsible for causing illnesses.

Hippocrates's most important contribution to medicine was arguably that he was one of the first doctors who used science to trace diseases to physical causes. Up to that time, most people believed that illnesses came from the gods or other supernatural agents. Doctors frequently treated illnesses using magic and sacrifices to their gods. Hippocrates's approach to medicine, however, brought important changes as he applied rational reasoning to the diagnosis and treatment of his patients. Not until Louis Pasteur came along in nineteenth-century France would the world fully understand that bacteria caused illnesses.

ARISTOTLE AND HEALTH

Aristotle's brilliant mind and keen understanding of a huge variety of topics is well documented. His father was a doctor, the court physician to the king of Macedon. Aristotle himself studied anatomy by dissecting animals. He explored biology by studying and cataloguing plants and animals. He is believed to be the first of the Greeks to use dissection as a tool for learning about anatomy. Several centuries later, Claudius Galenus—known more famously as Galen—was an important Greek doctor and surgeon. Galen's knowledge of medicine draws heavily from Aristotle's work. Where earlier Greeks had proposed the basic elements of earth, water, air, and fire, Aristotle proposed that there were four basic qualities: hot, cold, wet, dry. Aristotle is considered to be the father of comparative anatomy and physiology, which is the branch of biology that focuses on the ways organisms function.

CHAPTER SIX

AT HOME AND ON THE BATTLEFIELD: ANCIENT GREEKS AND ENGINEERING

Many of the Greeks' contributions in philosophy, science, mathematics, and medicine are well known. Their similarly impressive contributions in engineering are perhaps not so famous. Still, ancient Greece gave the world a variety of marvelous inventions. The Greek engineering inventions were often practical in nature, meant to solve a pressing problem.

To bring the water they needed to their cities, they built sophisticated, long-distance aqueducts. They also perfected the construction of cisterns, the large reservoirs that held a city's water supply. The Greeks invented a portable sundial and a water alarm clock (credited to Plato the philosopher). To grind grain more efficiently, the Greeks invented the water mill. The influences of Greek architecture are evident anywhere in the world today where there are buildings with columns or stadiums with rows of seats that rise from the ground up. Both concepts were architectural inventions of the Greeks.

The ancient Greeks were frequently at war, either fighting civil wars among themselves or fighting against other countries in an effort to expand their territory. Their invention of the catapult helped them—and others—win countless victories.

HIPPODAMUS OF MILETUS

In modern times, urban planners apply their skills in the design and planning of cities. Today's urban planners might trace their roots to the ancient Greek architect Hippodamus of Miletus (around 498–408 BCE), the first known city planner. Hippodamus developed the idea of building cities on a rectangular grid plan that resulted in city blocks. Before Hippodamus, many of the Greek cities sprang up without much planning, with haphazardly laid city streets that simply expanded outward as populations grew.

The concept of constructing city streets and buildings along perpendicular lines came to be called the Hippodamian grid or Hippodamian plan. Hippodamus went a step further than simply developing a practical, logical plan for the city streets. He also explored ideas about how a city could nurture the city residents, creating the best environment possible where they could thrive. Hippodamus applied his design ideas in a number of cities, including in his hometown of Miletus. There, he arranged the streets and the buildings so that the breezes blowing from the mountains and the sea flowed through the streets to cool the city in the hot summer months.

IKTINOS AND KALLIKRATES

When thinking about the engineering achievements of the ancient Greeks, it is important to remember that all they accomplished was done without the computer software programs and capabilities that are available to engineers working today. Instead, they had to rely on natural talent and ingenuity to create some of the ancient world's most impressive structures. Iktinos and Kallikrates were the two architects responsible for one of Greece's greatest edifices: the Parthenon. In Greece, architects and engineers built small-scale wooden prototypes before they began work on an actual building. They referred to the miniature wooden model as they worked to construct the real structure.

The Greeks built the Parthenon around 447 BCE as a temple to the goddess Athena. What remains of the Parthenon today—though mainly ruins—gives a hint of the simple, yet beautiful building it must have originally been. Its simple appearance, however, is deceiving—the Parthenon is based on complex math. The 9:4 ratio is used throughout the temple. Comparing the length of the building to the width results in a 9:4 ratio, which is also the ratio of the width of the front of the building compared to its height. When the distance between the columns—measuring from center to center of each column—is compared to the column diameter, the ratio is, again, 9:4. The consistent use of the ratio in construction results in a structure that is simple, yet elegant and beautiful, too.

The Parthenon, a masterpiece of ancient Greek architecture, sits atop the Acropolis of Athens.

Researchers have marveled at the precision, attention to detail, and craftmanship on display in the remains of the Parthenon. Without any of the machinery or skills available in modern times, the Greeks moved about 100,000 tons (90,718 metric tons) of marble from a quarry about 10 miles (16 km) away from the Parthenon to create this magnificent structure that was about half the size of an American football field at its base. Amazingly, all this work was completed in about eight or nine years. It is most likely that the Greeks used their advanced understanding of ropes, wooden cranes, and pulleys to accomplish the task.

The Parthenon, while damaged with the passing of time, still stands today. It has survived fires and earthquakes—a testimony to the skills of the ancient Greek engineers and architects. They had to overcome enormous construction problems, in addition to having to build their entire knowledge base from scratch. Despite everything being stacked against the architects, the Parthenon has been able to withstand earthquakes and other natural disasters for well over two thousand years.

ARCHITECTURE IN ANCIENT GREECE

Greek architects gave the world beautiful buildings, famous around the globe for their simple design, proportion, harmony, and perspective. Greek houses were commonly constructed of white stone with a courtyard in the center and a flat roof. Evidence of the influence of Greek architecture can been seen in many cities around the world today.

In Greek architecture, there were standards and rules about what columns should look like, especially in the temples of their gods. The Greeks called these rules orders. There were three main orders in their architecture: the Doric, Ionic, and Corinthian. Doric

(continued on the next page)

(continued from the previous page)

columns are simple in design. Ionic columns are more ornate and decorative. Corinthian columns were the most elaborate of the three, with very decorative features. The Parthenon in Athens displays Doric columns.

Architects followed order and logic in the design of their buildings. They even used advanced math, especially geometry, in their design calculations. This attention to detail resulted in buildings both sturdy and beautiful. Many examples of ancient Greek architecture are still standing today, in addition to the countless buildings inspired by their design and style.

The ancient Greeks were famous for their beautiful and intricate columns—pictured here are columns built in the Ionic order of architecture.

DIONYSIUS THE ELDER OF SYRACUSE

Greek engineering was useful for far more than just architecture and construction. Dionysius of Syracuse invented the catapult around 400 BCE. It changed warfare forever, giving the army that possessed catapults a definitive advantage over armies that did not. In its simplest form, the catapult is like a large, powerful slingshot. Soldiers attacked their enemies using catapults to send stones or other heavy objects over the walls of a city or castle. Some catapults were powerful, capable of hurling large stones across huge distances and over high walls.

The ancient Greeks invented the catapult, a weapon that gave an army a strong advantage over their enemies. A replica of one such contraption is shown here.

Modern researchers believe that Dionysius's earliest version of the catapult did not involve the use of any math in its creation. Craftsmen were able to construct these weapons even without a knowledge of geometry. When Archimedes came along—a little later than Dionysius—he fine-tuned the catapult using math theory to make the weapon. By applying geometry to an engineering problem, scientists made the catapult more precise in striking a target.

The concept of the catapult still has uses in modern-day life, even if its use in war has become obsolete. Many militaries launch airplanes from the short runways of aircraft carriers at sea using catapult action. In the classroom, teachers use catapults to demonstrate concepts in physics and math for their students.

INVENTIVE ARCHIMEDES

Archimedes, an astronomer and mathematician, was also an inventor. Legends exist surrounding some of the inventions and discoveries Archimedes made. One of the legends revolves around Archimedes searching for a solution to get rid of the rainwater that collected in the hulls of ships. Archimedes invented a machine that consisted of a hollow tube and a spiral with a handle on one end. By placing the tube into water and turning the handle, Archimedes could bring the water up the tube and out of the ship. This device, called an Archimedes screw, is still used in some developing countries around the world for water irrigation.

Another legend is that Archimedes was asked to determine whether or not an object was purely gold. A king was worried that the crown maker was using less valuable materials instead of real gold. The solution he concocted was to measure the crown's volume based on the water it displaced. Once he knew the volume of the crown and the crown's weight, he could calculate the crown's density and solve the question of whether or not it was pure gold. In solving this problem, he developed the law of buoyancy called Archimedes's principle today.

Archimedes was also able to make use of geometry to explain how devices like pulleys, fulcrums, and levers worked. He once reportedly used nothing but cogs and pulleys to move a fully loaded ship out of the water onto land, a job that normally required a hundred men.

Although there may be some mix of truth and fiction surrounding how he invented these devices, it is inarguable that these inventions both existed and changed the world. Archimedes has been compared to Leonardo da Vinci in his remarkable mechanical genius.

TIMELINE

620 BCE Thales, one of the first Western philosophers and mathematicians, is born.

610 BCE Anaximander, a pioneer cartographer, is born. He is one of the earliest mapmakers to create a world map.

570 BCE Pythagoras, a mathematician and musician, is born in Samos.

508 BCE Democracy is first seen in Athens.

490–480 BCE Persians invade Greece, leading to the Persian Wars. The Greeks are victorious, preserving their culture and influence in the world.

460 BCE Hippocrates, known as the father of medicine and the founder of the Hippocratic School of Medicine, is born.

447 BCE The Parthenon, temple to the goddess Athena, is constructed.

430 BCE Dionysius the Elder of Syracuse, who invented the catapult, is born.

384 BCE The great philosopher Aristotle is born.

336–323 BCE Alexander the Great rules Macedonia, spreading Greek culture, language, and learning across Asia Minor, Egypt, India, and Mesopotamia.

310 BCE Aristarchus, a mathematician and astronomer, is born.

300 BCE Euclid lives around this time, innovating mathematics and geometry.

287 BCE Archimedes, an inventor, astrologer, and mathematician, is born.

276 BCE Eratosthenes, who studied math, geography, astronomy, music, and poetry, is born.

GLOSSARY

anatomy The study of the parts of the human body.

archaic period The period in ancient Greek life from about 800 to 500 BCE.

Archimedes's principle The physical law of buoyancy, developed by Archimedes.

atom A building block of matter; first proposed by Democritus, who referred to them as *atoma*.

botany The study of plants.

calculus A branch of mathematics developed from geometry and algebra.

catapult A weapon developed by the Greeks; used to throw heavy objects at enemy combatants.

city-state A term describing the powerful cities and the small villages around them in ancient Greece.

classical period The period in ancient Greek life from about 500 BCE to 323 BCE. Famous philosophers Socrates, Plato, and Aristotle lived during this time.

democracy A form of government developed by the ancient Greeks in which the people have the power.

geocentric Referring to the belief that the earth is the center of the universe.

geometry The branch of mathematics, developed by Euclid, related to shapes and figures.

heliocentric Referring to the belief that the sun is the center of the universe.

Hellenistic period The period in ancient Greek life from about 323 BCE to 146 BCE when the Romans conquered Greece.

natural selection The idea that the organisms best suited to their environment will survive.

Parthenon The temple built in Athens around 447 BCE. It stands today, displaying the ancient Greeks' amazing engineering skills.

philosophy A way of thinking about the world, developed by the ancient Greeks.

Pythagorean theorem The math formula, developed by Pythagoras, that describes the relationships between the sides of a right triangle.

solstice The two times of the year, in June and December, when the sun is the greatest distance from the earth's equator.

STEM The acronym for science, technology, engineering, and mathematics.

zoology The study of animals.

FOR MORE INFORMATION

Metropolitan Museum of Art
Greek and Roman Art
1000 Fifth Avenue
New York, NY 10028
(212) 535-7710
Website: https://www.metmuseum.org/about-the-met/curatorial-departments/greek-and-roman-art
Facebook, Instagram, and Twitter: @metmuseum
This New York City art museum has an extensive collection of ancient Greek art. Highlights of the collection are available on its website.

Museum of Antiquities
University of Saskatchewan
Peter Mckinnon Building
107 Administration Place
University of Saskatchewan
Saskatoon, SK S7N 5A2
Canada
(306) 966-7818
Website: https://artsandscience.usask.ca/antiquities
Facebook: @usaskantiquities
Instagram and Twitter: @MofAntiquities
This Saskatoon, Canada, archaeological museum features ancient Greek sculptures and art.

National Hellenic Museum
333 South Halsted Street
Chicago, IL 60661
(312) 655-1234
Website: https://www.nationalhellenicmuseum.org
Facebook: @NationalHellenicMuseum
Instagram and Twitter: @HellenicMuseum
This Chicago museum displays the cultural contributions of ancient and contemporary Greeks.

Royal Ontario Museum
100 Queen's Park
Toronto, ON M5S 2C6
Canada
(416) 586-8000
Website: https://www.rom.on.ca/en
Facebook: @royalontariomuseum
Instagram and Twitter: @ROMtoronto
This Toronto museum features the Gallery of Greece, displaying Greek history from the archaic and classical periods.

Science History Institute
315 Chestnut Street
Philadelphia, PA 19106
(215) 925-2222
Website: https://www.sciencehistory.org
Facebook, Instagram, and Twitter: @SciHistoryOrg
The institute's museum contains a wide-ranging collection documenting the history of science.

University of Pennsylvania Museum of Archaeology and Anthropology
3260 South Street
Philadelphia, PA 19104
(215) 898-4000
Website: https://www.penn.museum
Facebook, Instagram, and Twitter: @pennmuseum
This museum has a Greece Gallery featuring ancient Greek history and culture. The Ancient Greek World virtual gallery is available on its website.

FOR FURTHER READING

Bauer, Susan Wise. *The Story of Western Science: From the Writings of Aristotle to the Big Bang Theory*. New York, NY: W. W. Norton & Company, 2015.

Bordessa, Kris. *Tools of the Ancient Greeks: A Kid's Guide to the History & Science of Life in Ancient Greece*. White River Junction, VT: Nomad Press, 2006.

Cottrell, George. *Ancient Greece*. New York, NY: Kidhaven Publishing, 2017.

Ede, Andrew, and Lesley B. Cormack. *A History of Science in Society, Volume 1: From the Ancient Greeks to the Scientific Revolution*. Toronto, Canada: University of Toronto Press, 2017.

Evans, Maz. *Who Let the Gods Out?* Somerset, UK: Chicken House, 2017.

Fagan, Brian M. *The Seventy Great Inventions of the Ancient World*. New York, NY: Thames & Hudson, 2004.

Fullman, Joseph. *Ancient Civilizations: Discover the Golden Ages of History, from Ancient Egypt and Greece to Mighty Rome and Exotic Maya*. New York, NY: DK Eyewitness Books, 2013.

Gerster, Georg, and Paul Cartledge. *The Sites of Ancient Greece*. New York, NY: Phaidon Press, 2012.

Hill, Stuart. *Ancient Wonders—Then & Now*. New York, NY: Lonely Planet Kids, 2018.

Machajewski, Sara. *A Kid's Life in Ancient Greece*. New York, NY: PowerKids Press, 2015.

Pearson, Anne. *DK Eyewitness Books: Ancient Greece*. New York, NY: DK Children, 2014.

BIBLIOGRAPHY

Adhikari, Saugat. "Top 10 Inventions and Discoveries of Ancient Greeks That Are Remarkably Used Today." Ancient History Lists, February 11, 2019. https://www.ancienthistorylists.com/greek-history/top-10-inventions-discoveries-ancient-greece-remarkably-used-today.

Allen, Don. "The Origins of Greek Mathematics." Texas A&M University, January 13, 1997. http://www.math.tamu.edu/~dallen/history/greekorg/greekorg.html.

Brazier, Yvette. "What Is Ancient Greek Medicine?" Medical News Today, November 6, 2018. https://www.medicalnewstoday.com/articles/323596.php.

British Museum. "Ancient Greece." Retrieved March 7, 2019. http://www.ancientgreece.co.uk.

Canadian Museum of History. "Science and Philosophy." Retrieved March 7, 2019. https://www.historymuseum.ca/cmc/exhibitions/civil/greece/gr1080e.html.

Carr, K. E. "Ancient Greek Science—Math, Astronomy, Medicine." Quatrus Study Guides, July 18, 2017. https://quatr.us/greeks/ancient-greek-science.htm.

Cartwright, Mark. "Ancient Greek Inventions." Ancient History Encyclopedia, December 20, 2017. https://www.ancient.eu/article/1165/ancient-greek-inventions.

Department of Classics. "Ancient Philosophy and Science." University of Pittsburgh. Retrieved March 5, 2019. https://www.classics.pitt.edu/research/ancient-philosophy-and-science.

Hadingham, Evan. "Unlocking Mysteries of the Parthenon." *Smithsonian*, February 2008. https://www.smithsonianmag.com/history/unlocking-mysteries-of-the-parthenon-16621015.

Harold B. Lee Library. "Greece: Ancient and Classical." Brigham Young University. Retrieved February 23, 2019. https://eudocs.lib.byu.edu/index.php/Greece:_Ancient_and_Classical.

Leafloor, Liz. "Incredible Construction: Greek Acropolis Built by Ancient Engineers to Resist Earthquakes." Ancient Origins, March 25, 2015. https://www.ancient-origins.net/news-history-archaeology/incredible-construction-greek-acropolis-ancient-earthquakes-020171.

Mark, Joshua J. "Ancient Greece." Ancient History Encyclopedia, November 13, 2013. https://www.ancient.eu/greece.

Martin, Rebecca. "Ancient Greek Cisterns." The Civil Engineer. Retrieved March 26, 2019. https://www.thecivilengineer.org/online-historical-database-of-civil-infrastructure/item/384-ancient-greek-cisterns.

National Geographic. "Science and Curiosity in Ancient Greece." Retrieved March 29, 2019. https://www.nationalgeographic.org/media/science-and-curiosity-ancient-greece.

News Staff. "Engineering Without Complex Math: Ancient Greek Craftsman and Their Magnificent Machines." Science 2.0, October 2, 2007. https://www.science20.com.

TimeMaps. "Ancient Greek Civilization." TimeMaps. Retrieved March 26, 2019. https://www.timemaps.com/civilizations/ancient-greeks.

Violatti, Cristian. "Ancient Greek Science." Ancient History Encyclopedia, June 13, 2013. https://www.ancient.eu/Greek_Science.

Walsh, David. "Top 4 Ancient Design Prototypes." American Society of Mechanical Engineers, August 2015. https://www.asme.org/engineering-topics/articles/manufacturing-design/top-4-ancient-design-prototypes.

INDEX

N

O

P

R

S

T

W

Z

ABOUT THE AUTHOR

Donna B. McKinney is a writer who lives in North Carolina. She spent many years writing about science and technology topics at the US Naval Research Laboratory in Washington, DC. Now she enjoys telling stories about topics ranging from science to history to sports for children and young adults. She has visited Athens and Corinth, where evidence of the flourishing culture of the ancient Greeks is still visible today.

PHOTO CREDITS

Cover Lambros Kazan/Shuttertstock.com; p. 5 Museo Archeologico e d'Arte Della Maremma, Grosseto, Italy/De Agostini Picture Library/S. Vannini/Bridgeman Images; p. 8 Perry-Castañeda Library Map Collection/University of Texas Libraries, The University of Texas at Austin; p. 10 Planet Observer/Universal Images Group /Getty Images; p. 13 Heritage Image/Hulton Archive/Getty Images; p. 18 Hulton Archive/Getty Images; pp. 20, 44, 51 DEA/G. Dagli Orti/De Agostini/Getty Images; p. 23 Print Collector/Hulton Archive/Getty Images; p 27 Universal History Archive/Universal Images Group/Getty Images; p. 29 De Agostini/G. Cigolini/De Agostini Picture Library/Getty Images; p. 33 Jay M. Pasachoff /Science Faction/Getty Images; p. 35 Science Source/Getty Images; p. 38 Universal Images Group/Getty Images; p. 42 Stock Montage /Archive Photos/Getty Images; p. 49 Zwiebackesser/Shutterstock .com; p. 53 DEA/A. Dagli Orti/De Agostini/Getty Images; p. 60 Guy Vanderelst/Photolibrary/Getty Images; p.62 PHAS/Universal Images Group/Getty Images; p. 63 Xinhua/Alamy Stock Photo; cover and interior pages (dark textured background) Midiwaves /Shutterstock.com; interior pages (scroll pattern page borders) Megin/Shutterstock.com, (yellow marbled page borders) Chizhovao/Shutterstock.

Design and Layout: Nicole Russo-Duca; Editor: Siyavush Saidian; Photo Researcher: Cindy Reiman